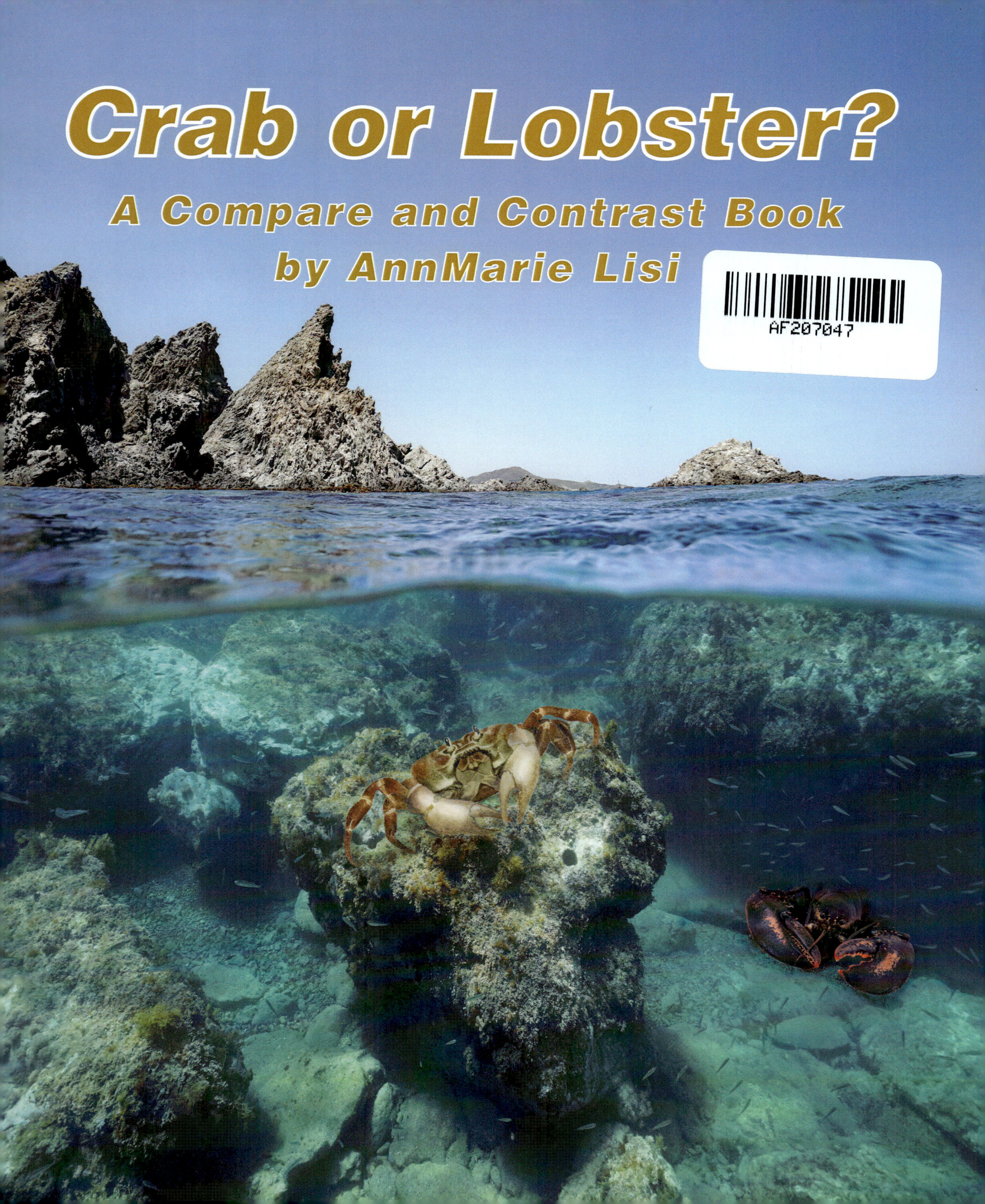

Crab or Lobster?
A Compare and Contrast Book
by AnnMarie Lisi
AF207047

Crabs and lobsters are water-living (aquatic) crustaceans.

Unlike mammals, birds, fish, amphibians, or reptiles, crustaceans do not have a backbone—they are invertebrates. Both have skeletons on the outside of their bodies called exoskeletons. Like insects, crabs and lobsters have segmented bodies and jointed legs.

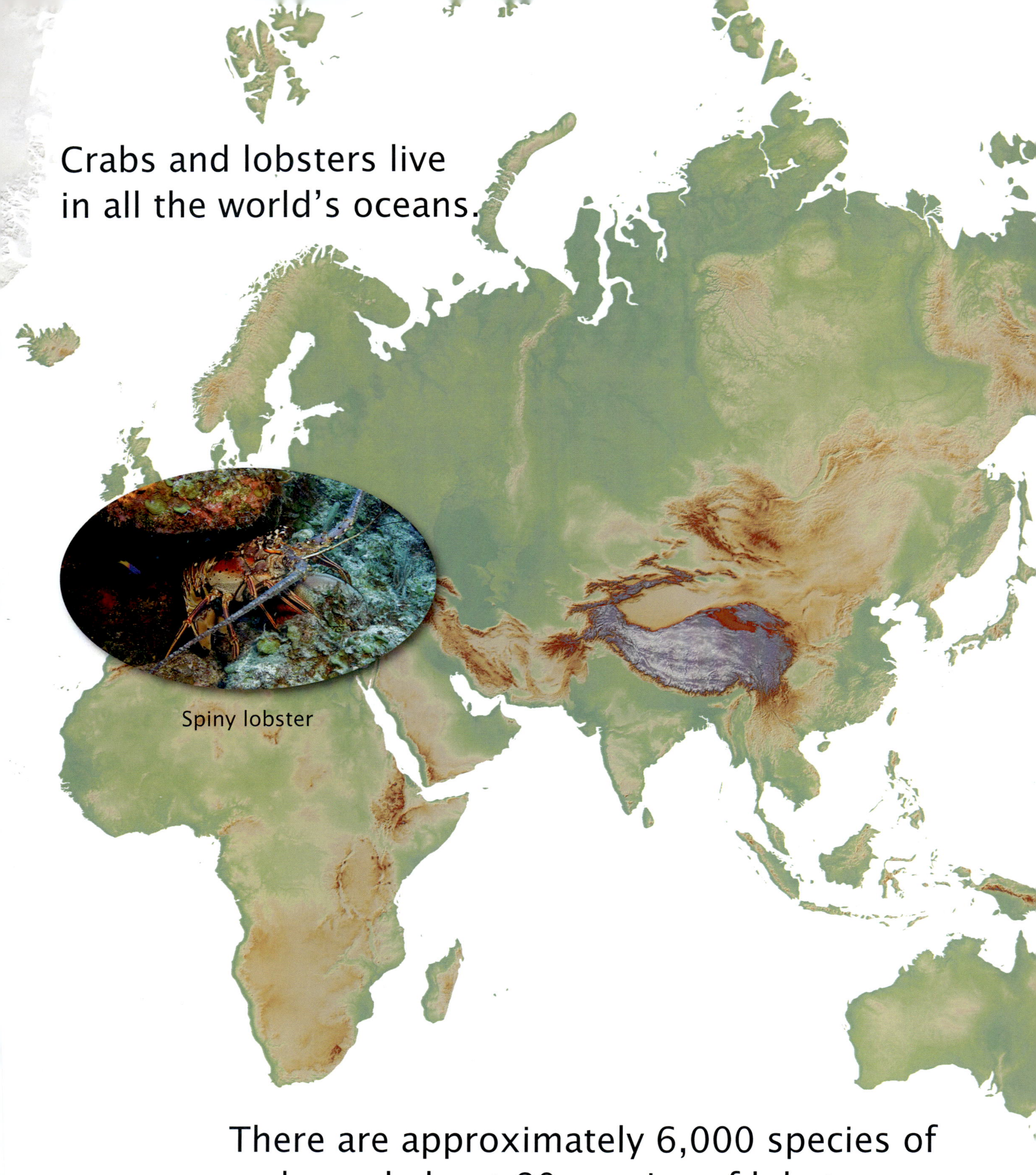
Crabs and lobsters live
in all the world's oceans.

Spiny lobster

There are approximately 6,000 species of
crabs and about 80 species of lobsters.

Alaskan king crabs live in cold, Arctic waters.

New England lobsters live in the more temperate waters of the North Atlantic.

Spiny lobsters live in warm-water coral reefs in the Mediterranean and Caribbean Seas.

Rocky shorelines with tide pools are a great place to look for crabs.

Lobsters live near the coast, but you wouldn't usually find them in a tide pool.

Their hard exoskeletons protect them from predators like large fish or seals.

They also hide between rocks and blend (camouflage) into their surroundings for protection.

Can you find the crabs and lobsters?

When crabs hatch, they have a tiny tail to help them move through the water. As they grow up, their tail folds underneath their bodies.

As humans grow, our bones grow with us. Many animals can't do that. Just like snakes shed their skin as they grow, both crabs and lobsters grow by molting or sliding out of their exoskeletons.

It takes a few days for their new exoskeleton to become hard. During that time, they usually hide so they don't become an easy meal for predators.

Sometimes you find these molts washed up on the beach. Sometimes a crab or lobster will eat their molt!

crab larva

spiny lobster larva

crab molt

lobster molt

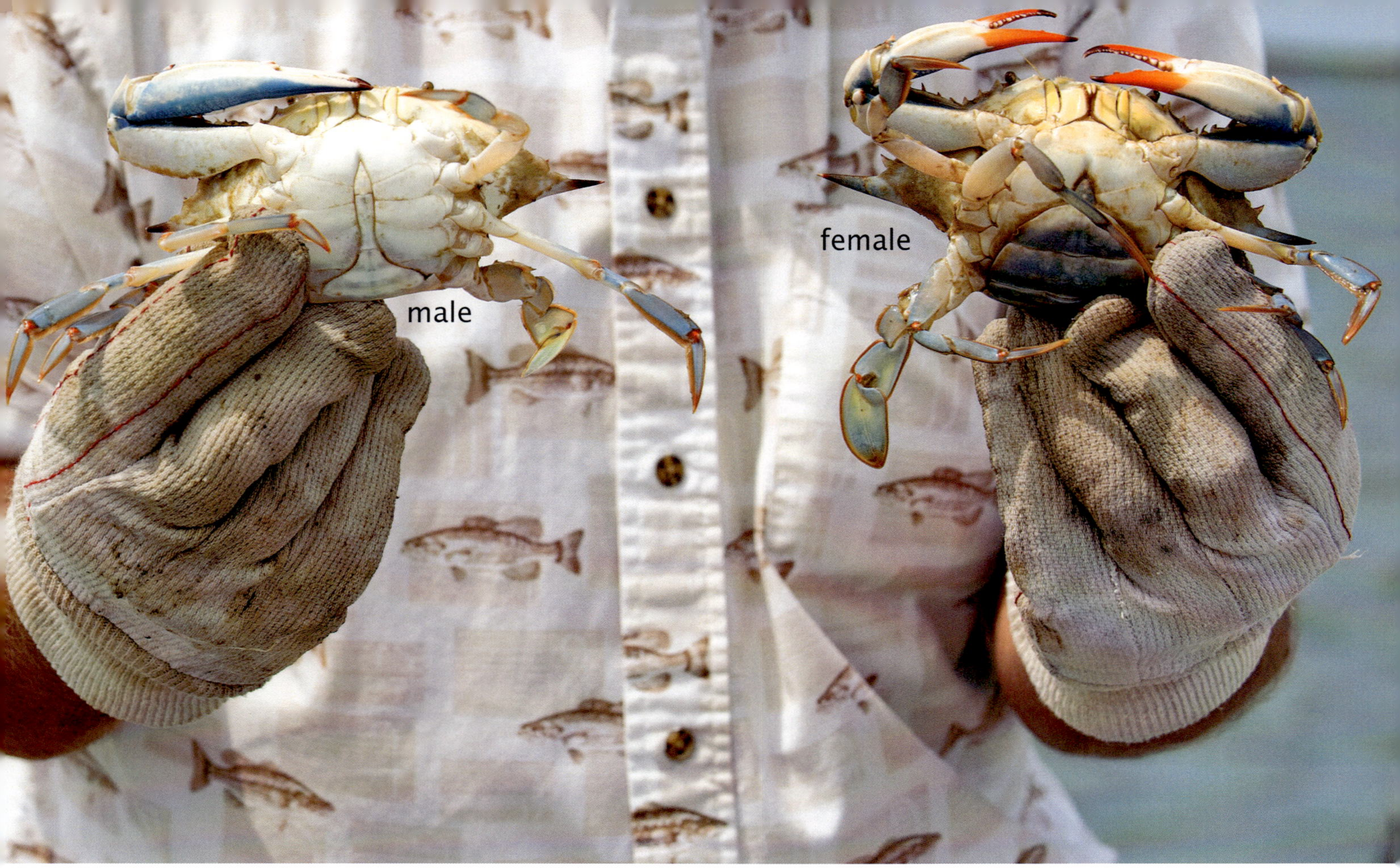

If you look at the underside of a crab you can tell a male from a female by looking at the tail.

Males usually have a very narrow and pointy tail.

Females have a large, dome-shaped tail to store eggs.

Female lobsters store eggs under their tails, too.

Crabs can live
in salt water,
fresh water, or
even on land!

The coconut crab lives
on land. It is the largest
crustacean and can grow
up to 3 feet (1 m) wide.

The pea crab is the smallest crab at ½ inch (2.5 cm) wide. This crab is a parasite that lives inside of oysters and other shellfish.

Lobsters live in salt
water. They can be
found in rocky or
muddy areas near
the shore.

Sometimes, crayfish are mistaken for lobsters. Even though the two are related, crayfish live in freshwater lakes, rivers, and streams and are smaller than lobsters.

Most crabs' claws are the same size and shape.

Male fiddler crabs are the exception. They have one large claw and one small claw. They get their name "fiddler" because their large claw looks like a violin. They use the large claw to attract a mate.

Lobsters have ten legs.

Look closely at these lobsters. You'll notice their front legs are claws with two different shapes. The long pointy claw is called the pincher, and the large fat claw is called the crusher.

Now look at these spiny lobsters and you'll notice they do not have any large claws!

In addition to claws, crabs have legs for walking. Some crab species have swimmerets, or paddles, on their back legs to swim through the water.

Lobsters walk along the ocean floor. They can quickly use their tail to push water and scoot in a backward direction.

Both crabs and lobsters have their eyes at the top of slender appendages called "eyestalks" that extend from their bodies. Having their eyes on stalks allows them to easily see in any direction, even behind them.

Crabs and lobsters have two sets of antennae. One set is used for feeling around its habitat. The other is used for smelling in the water to help them find food.

Crab Body Parts

Lobster Body Parts

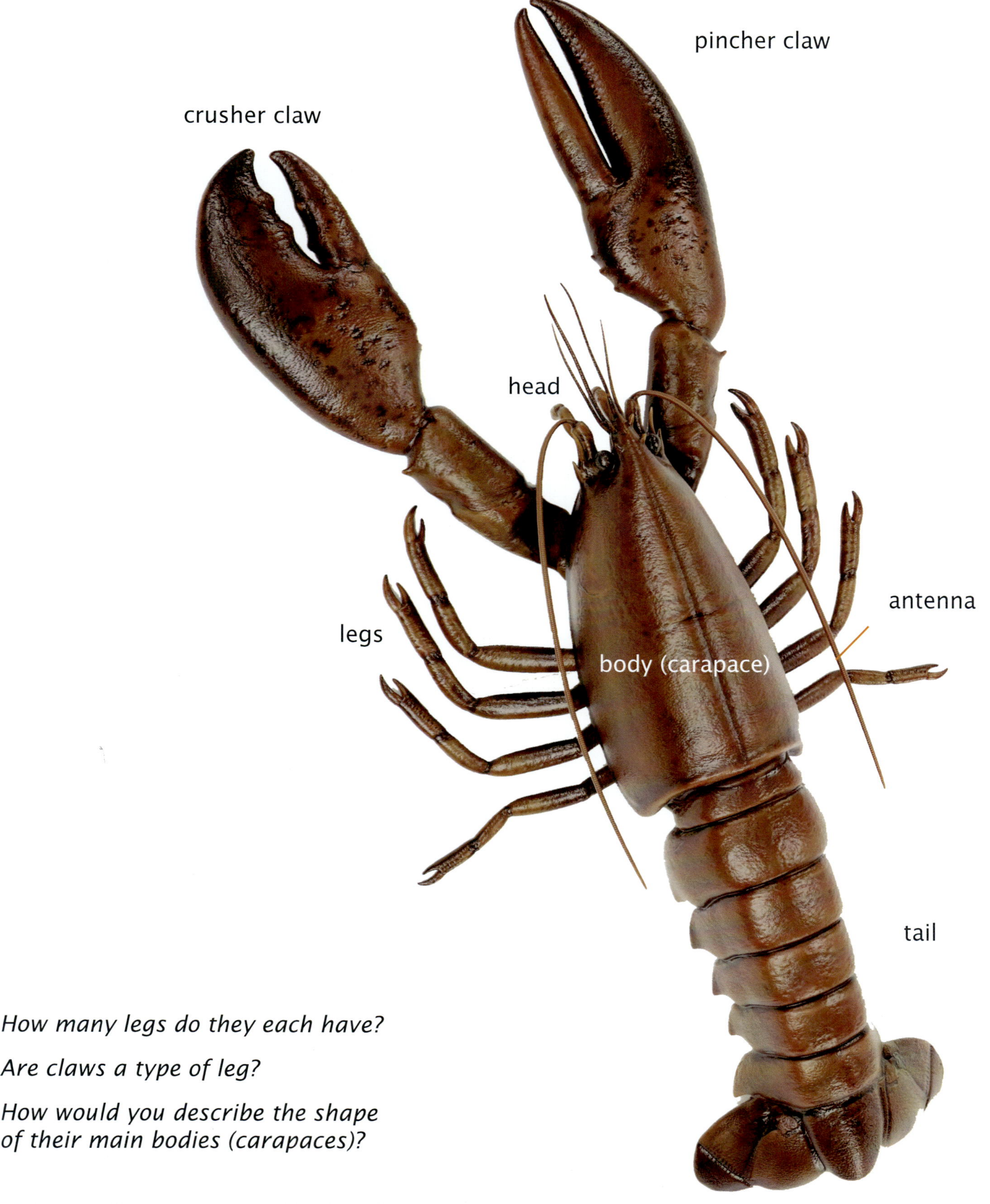

How many legs do they each have?

Are claws a type of leg?

How would you describe the shape of their main bodies (carapaces)?

Classifying Crabs and Lobsters

Within the Animal Kingdom, animals are initially divided into phyla. Within each phylum, the animals are then sorted into classes, then into orders, suborders, families and finally into a genus and species. No matter what language scientists speak, they use animals' genus and species to identify specific animals. Those names are always in Latin.

Let's see how crabs and lobsters are classified.

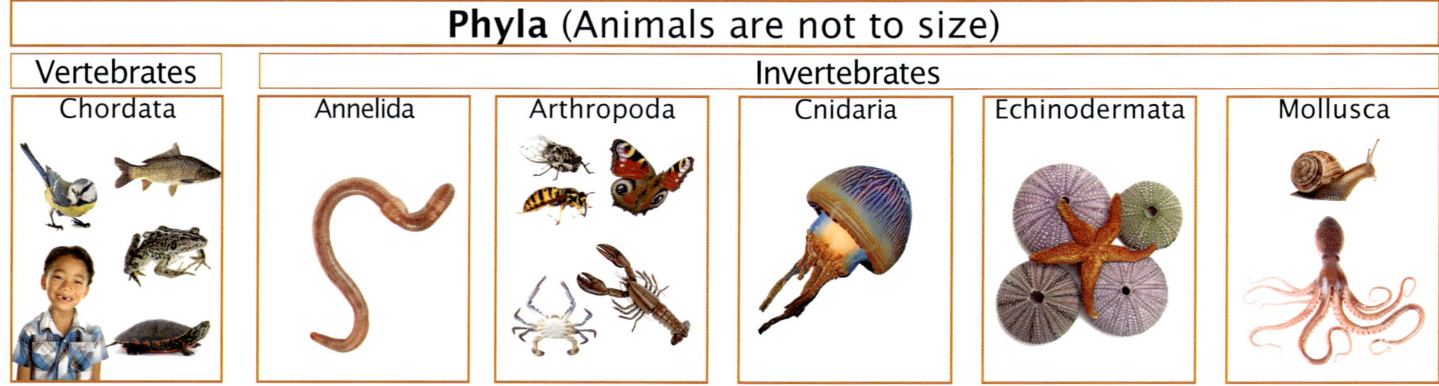

You may have already studied and learned about the five classes of animals that have backbones (Vertebrates/Chordata): mammals, fish, birds, reptiles, and amphibians.

Did you know that there are actually more animals on the earth that do NOT have backbones? They are called invertebrates. Just like the five vertebrate classes, the invertebrate phylum is also broken into classes as shown above.

Crabs and lobsters (Crustaceans) are Arthropods (Arthropda class). Other arthropods include spiders (Arachnids), centipedes and millipedes (Myriapoda), and insects (Hexapoda).

Crustaceans mostly live in the water (aquatic). They have segmented bodies and two pairs of antennae. They have a hard-shell skeleton on the outside of their bodies (exoskeleton) that they shed with a new one growing underneath (molt) as they grow.

Match the Arthropod

Can you identify which animals belong to the Arthropod classes?

Animals shown are not to size.

Arachnid: air-breathing animals with four pairs of legs (spiders & scorpions)
Crustacean: aquatic animals with segmented bodies, antennae, and exoskeleton
Hexopoda: animal with six legs and may have wings (insects)
Myriapoda: animal with a long body and similar segments (centipedes & millipedes)

beetle	black widow spider	blue crab	butterfly	centipede

dragonfly	fiddler crab	fly	hermit crab	lobster

millepede	spiny lobster	shrimp	wasp	wolf spider
				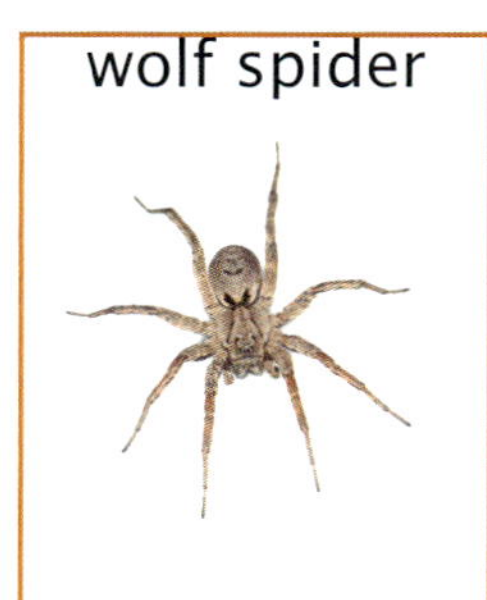

Note to adult: have children explain why they think the animal fits which group.

Answers:
Arachnid: black widow spider, wolf spider
Crustacean: blue crab, fiddler crab, hermit crab, lobster, spiny lobster, shrimp
Hexopoda: beetle, butterfly, dragonfly, fly, wasp
Myriapoda: centipede, millepede

Library of Congress Cataloging-in-Publication Data

Names: Lisi, AnnMarie, 1985- author.
Title: Crab or lobster? : a compare and contrast book / by AnnMarie Lisi.
Description: Mt. Pleasant, SC : Arbordale Publishing, [2023] | Series:
   Compare and contrast | Includes bibliographical references.
Identifiers: LCCN 2023054073 (print) | LCCN 2023054074 (ebook) | ISBN
   9781643519906 (English paperback) | ISBN 9781638170099 (dual-language,
   read along) | ISBN 9781638170280 (pdf) | ISBN 9781638170471 (epub)
Subjects: LCSH: Crabs--Juvenile literature. | Lobsters--Juvenile
   literature. | Decapoda (Crustacea)--Juvenile literature.
Classification: LCC QL444.M33 L568 2023 (print) | LCC QL444.M33 (ebook) |
   DDC 595.3/84--dc23/eng/20231229
LC record available at https://lccn.loc.gov/2023054073
LC ebook record available at https://lccn.loc.gov/2023054074

Also available in Spanish: *¿Cangrejo o langosta? Un libro de comparaciones y contrastes*
Spanish Paperback   9781638172918
Spanish PDF   9781638172994
Spanish ePub3 9781638173038
The dual-language read-along is available online at www.fathomreads.com

Bibliography

A-Z-Animals.com. "Animal Classification." A-z-Animals.com, 2018, a-z-animals.com/reference/animal-
   classification/.
"ADW: Arthropoda: CLASSIFICATION." Animaldiversity.org, animaldiversity.org/accounts/Arthropoda/
   classification/#Arthropoda.
"List of Crustaceans | Britannica." Www.britannica.com, www.britannica.com/topic/list-of-crustaceans-2034273.

Printed in the US
This product conforms to CPSIA 2008

Text Copyright 2024 © by AnnMarie Lisi

The "For Creative Minds" educational section may be
copied by the owner for personal use or by educators
using copies in classroom settings.

Arbordale Publishing, LLC
Mt. Pleasant, SC 29464
www.ArbordalePublishing.com